COMMON CORE CLINICS

Grade 1

Mathematics

CLINICS

Operations and Algebraic Thinking

Common Core Clinics, Mathematics, Operations and Algebraic Thinking, Grade 1
OT281 / 473NA

ISBN: 978-0-7836-8583-0

Contributing Writer: Andie Liao
With special thanks to mathematics consultants:
Debra Harley, Director of Math/Science K-12, East Meadow School District
Allan Brimer, Math Specialist, New Visions School, Freeport School District
Cover Image: © Alloy Photography/Veer

Triumph Learning® 136 Madison Avenue, 7th Floor, New York, NY 10016

© 2012 Triumph Learning, LLC
Coach is an imprint of Triumph Learning®

ALL ABOUT YOUR BOOK

COMMON CORE CLINICS will help you master important math skills.

Get the Idea shows you how to solve problems.

Guided Practice gives you help while you practice.

A **Glossary** and **Math Tools** will help you work out problems.

Try It! lets you practice on your own.

Table of Contents

Fact Families

Get the Idea

A **fact family** is a group of related facts.
The facts use the same numbers.

4, 6, and 10 are in the fact family below.

related addition facts

$6 + 4 = 10$

$4 + 6 = 10$

related subtraction facts

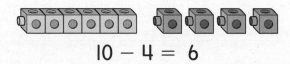

$10 - 4 = 6$

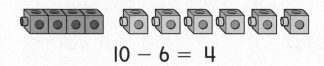

$10 - 6 = 4$

Some fact families have only 2 facts.
Think about doubles.

5, 5, and 10 are in this fact family.

$5 + 5 = 10$

$10 - 5 = 5$

Example 1

What are the related subtraction facts?

$7 + 6 = 13$

Strategy Use the same numbers.

Step 1 What are the numbers in the fact?
$6, 7, 13$

Step 2 Write the related facts.
Use the same numbers.

$$13 - 6 = 7$$
$$13 - 7 = 6$$

Answer The related subtraction facts are:
$13 - 6 = 7$
$13 - 7 = 6$

Example 2

Write a fact family with these numbers.

6 8 14

Strategy Use the same numbers.

Step 1 Write the addition facts.

$$8 + 6 = 14$$
$$6 + 8 = 14$$

Step 2 Write the subtraction facts.

$$14 - 8 = 6$$
$$14 - 6 = 8$$

Answer
$$8 + 6 = 14$$
$$6 + 8 = 14$$
$$14 - 8 = 6$$
$$14 - 6 = 8$$

Guided Practice

Write a fact family with these numbers.

5   7   12

What are the numbers in this fact family? _____

Write the addition facts.

____5____ + ____7____ = ____12____

_____ + _____ = _____

Write the subtraction facts.

____12____ – ____7____ = ____5____

_____ – _____ = _____

The fact family is:

_____ + _____ = _____

_____ + _____ = _____

_____ – _____ = _____

_____ – _____ = _____

Try It!

Fill in the ☐ for each fact family.

1. $3 + 8 = \boxed{}$

    $8 + 3 = \boxed{}$

    $\boxed{} - 8 = 3$

    $\boxed{} - 3 = 8$

2. $\boxed{} + 3 = 10$

    $3 + \boxed{} = 10$

    $10 - 3 = \boxed{}$

    $10 - \boxed{} = 3$

3. $\boxed{} + 5 = 13$

    $5 + \boxed{} = 13$

    $13 - 5 = \boxed{}$

    $13 - \boxed{} = 5$

4. $5 + 4 = \boxed{}$

    $4 + 5 = \boxed{}$

    $\boxed{} - 5 = 4$

    $\boxed{} - 4 = 5$

5. $4 + \boxed{} = 8$

    $8 - \boxed{} = 4$

6. $7 + \boxed{} = 14$

    $14 - \boxed{} = 7$

Operations and Algebraic Thinking

7. Peter uses these numbers.

    7   8   15

    Write the fact family.

    _____

    _____

    _____

    _____

    Write your own fact family.
    Use different numbers.

    _____

    _____

    _____

    _____

Get the Idea

You can add three numbers in any order.
Add two numbers first.

$$\boxed{5} + \boxed{3} + 2 =$$
$$\boxed{8} + 2 = 10$$

$$5 + \boxed{3} + \boxed{2} =$$
$$5 + \boxed{5} = 10$$

$$\boxed{5} + 3 + \boxed{2} =$$
$$3 + \boxed{7} = 10$$

Look for doubles or make a 10.

Example 1

```
    6
    2
+   6
───────
```

Strategy Look for doubles.

Step 1 Add 6 + 6.

```
 ┌───┐
 │ 6 │ ──────┐
 └───┘       ┌─────┐
    2        │ 12  │
 ┌───┐ ──────└─────┘
+│ 6 │       +   2
 └───┘      ───────
───────
```

Step 2 Add the third number.

```
   12
+   2
──────
   14
```

Answer

```
    6
    2
+   6
──────
   14
```

Example 2

$$1 + 7 + 3 = \boxed{}$$

Strategy Make a 10.

Step 1 Add 7 + 3.

$$1 + \boxed{7} + \boxed{3} =$$

$$1 + \boxed{10} =$$

Step 2 Add the third number.

$$1 + 10 = 11$$

Answer 1 + 7 + 3 = 11

Guided Practice

$8 + 2 + 3 = \boxed{}$

Add the numbers in any order.
Make a ten.

$8 + 2 =$ _____

Add the third number.

$10 + 3 =$ _____

$8 + 2 + 3 =$ _____

Try It!

1.
$$\begin{array}{r} 3 \\ 6 \\ + \ 2 \\ \hline \end{array}$$

2.
$$\begin{array}{r} 8 \\ 1 \\ + \ 5 \\ \hline \end{array}$$

3.
$$\begin{array}{r} 7 \\ 3 \\ + \ 2 \\ \hline \end{array}$$

4.
$$\begin{array}{r} 8 \\ 1 \\ + \ 8 \\ \hline \end{array}$$

5.
$$\begin{array}{r} 1 \\ 5 \\ + \ 9 \\ \hline \end{array}$$

6.
$$\begin{array}{r} 7 \\ 7 \\ + \ 3 \\ \hline \end{array}$$

7. $5 + 4 + 5 = $ _____

8. $8 + 7 + 2 = $ _____

9. Show two ways to add.

$4 + 4 + 6 = \boxed{}$

Use doubles.

Make a ten.

Lesson 3 Equations

Get the Idea

This number sentence is an **equation**.
It has an **equal sign (=)**.
Both sides equal 8.

$$5 + 3 = 4 + 4$$

Example 1

Is this equation true?

$7 + 2 = 9$

Strategy Model the numbers.

Step 1 Show $7 + 2$.

Step 2 Count how many 🏈 in all. There are 9 🏈 in all.

Step 3 Is the equation true? Yes, $7 + 2 = 9$.

Answer Yes, $7 + 2 = 9$ is true.

Example 2

Which is true?

$$5 + 5 = 9 \qquad 11 = 10 \qquad 12 - 3 = 9$$

Strategy Test each equation.

Step 1 Check $5 + 5 = 9$.

$$5 + 5 = 10$$

This equation is **not** true.

Step 2 Check $11 = 10$.

11 does not equal 10.

This equation is **not** true.

Step 3 Check $12 - 3 = 9$.

$$12 - 3 = 9$$

This equation is true.

Answer The equation $12 - 3 = 9$ is true.

Guided Practice

Which is true?

$$15 = 51 \qquad 7 + 6 = 16 \qquad 20 - 10 = 10$$

⬭ ⬭ ⬭

Test each equation.

Check $15 = 51$.

15 does __not__ equal 51.

This equation is _____ true.

Check $7 + 6 = 16$.

$7 + 6 = \underline{13}$

This equation is _____ true.

Check $20 - 10 = 10$.

$20 - 10 = \underline{}$

This equation _____ true.

The equation _____ is true.

Try It!

Choose the correct answer.

1. Which makes this equation true?

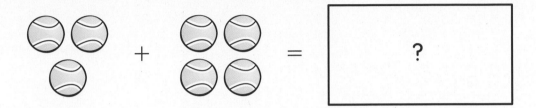

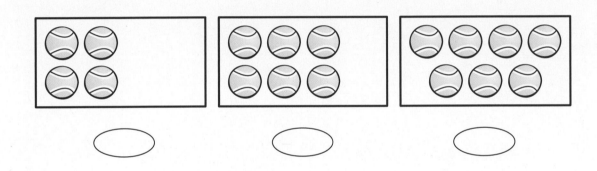

2. Which does this picture show?

$5 + 1 = 5 + 5$ $3 + 2 = 1 + 4$ $2 + 2 = 1 + 3$

3. **Which makes the equation true?**

    6 + 3 = ____ + ____

    5 + 4   2 + 8   7 + 4
    ◯   ◯   ◯

4. **Which makes the equation true?**

    2 + 9 = ____ + ____

    8 + 5   3 + 8   6 + 6
    ◯   ◯   ◯

5. **Which equation is true?**

    13 − 4 = 7   10 − 6 = 4   11 − 7 = 6
    ◯   ◯   ◯

6. **Which equation is true?**

    9 − 3 = 12   8 − 5 = 13   7 + 8 = 15
    ◯   ◯   ◯

7. Joe uses to show an equation.

Make the equation true. Draw the .

Write the equation.

_____ + _____ = _____

Lesson 4 Missing Numbers in Equations

Get the Idea

You can find a missing number in an equation.

$$9 + \boxed{} = 12$$

Start with 9.
Count on to 12.

$$9 + \boxed{3} = 12$$

Example 1

Find the missing number.

$\boxed{} + 6 = 11$

Strategy Use ◯. Count on.

Step 1 Start with 6.

Step 2 Count on to 11.

6 ◯ ◯ ◯ ◯ ◯
   7 8 9 10 11

Step 3 You counted on 5.

$\boxed{5} + 6 = 11$

Answer $\boxed{5} + 6 = 11$

You can use a related fact to find a missing number.

$9 + \boxed{} = 12$

$12 - 9 = \boxed{}$ is a related fact.

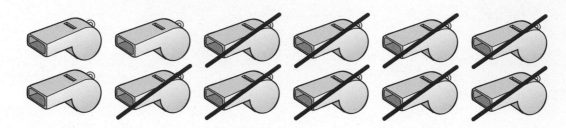

$9 + \boxed{3} = 12$

Example 2

Find the missing number.

$5 + \boxed{} = 13$

Strategy Use a related fact.

Step 1 Think about subtraction.

$13 - 5 = \boxed{?}$

Step 2 Use the same numbers.

$13 - 5 = \boxed{8}$

$5 + \boxed{8} = 13$

Answer $5 + \boxed{8} = 13$

Example 3

Find the missing number.

$$10 - \boxed{} = 7$$

Strategy Use a related fact.

Step 1 Think about addition.

$$7 + \boxed{?} = 10$$

Step 2 Use the same numbers.

$$7 + \boxed{3} = 10$$

$$10 - \boxed{3} = 7$$

Answer $10 - \boxed{3} = 7$

Guided Practice

Find the missing number.

$$\boxed{} - 9 = 4$$

Use a related fact.
Think about addition.

$$9 + 4 = \boxed{?}$$

Use the same numbers.

$$9 + 4 = \boxed{}$$

$$\boxed{} - 9 = 4$$

The missing number is _____.

$$\boxed{} - 9 = 4$$

Operations and Algebraic Thinking

Try It!

Choose the correct answer.

1. What is the missing number?

   $7 + \boxed{} = 11$

   7   8   9   10   11

   4   7   8

   ⬭   ⬭   ⬭

2. Which can you use to find ?

   $3 + \boxed{} = 8$

   5 − 3 = 2   8 − 3 = 5   5 − 2 = 3

   ⬭   ⬭   ⬭

3. Which can you use to find $\boxed{}$?

   $\boxed{} + 6 = 13$

   17 − 6 = 11   13 − 6 = 7   13 − 3 = 10

   ⬭   ⬭   ⬭

4. Which can you use to find $\boxed{}$?

$7 - \boxed{} = 4$

4 + 7 = 11 ⬭   4 + 3 = 7 ⬭   7 + 3 = 10 ⬭

5. Which can you use to find $\boxed{}$?

$\boxed{} - 5 = 5$

5 + 0 = 5 ⬭   6 + 6 = 12 ⬭   5 + 5 = 10 ⬭

6. What is the missing number?

$8 + \boxed{} = 14$

4 ⬭   6 ⬭   8 ⬭

7. What is the missing number?

$12 - \boxed{} = 7$

3 ⬭   4 ⬭   5 ⬭

8. **What is the missing number?**

     $- 8 = 4$

    10   11   12

    $\bigcirc$ $\bigcirc$ $\bigcirc$

9. **Ben writes this equation.**

     $+ 7 = 14$

    What is the missing number? _____

    Tell how you got the missing number.

    _____

    _____

    _____

    _____

    _____

    _____

Addition and Subtraction Word Problems

Get the Idea

You can solve a word problem in many ways.
Write an equation for the problem.

Example 1

8 fish are swimming. Some fish join them.

Now there are 17 fish. How many fish joined?

Strategy Write an equation.

Step 1 What do you need to find? how many fish joined

Step 2 Do you add or subtract? "Join" means to add.

Step 3 Write an equation. Add.

$$8 + \boxed{?} = 17$$
$$8 + \boxed{9} = 17$$

Answer 9 fish joined.

Example 2

There are 10 .

Tina hits 4 🎳. How many 🎳 are left?

Strategy Write an equation.

Step 1 What do you need to find?
how many 🎳 are left

Step 2 Do you add or subtract?
"How many left" means to subtract.

Step 3 Write an equation. Subtract.

$$10 - 4 = \boxed{?}$$
$$10 - 4 = \boxed{6}$$

Answer 6 🎳 are left.

Guided Practice

2 are in the . 4 are in the .

There are 9 in all.

How many are in the ?

Write an equation.

$$2 + 4 + \boxed{?} = 9$$

Add the first two numbers.

$$2 + 4 + \boxed{?} = 9$$

$$\underline{} + \boxed{?} = 9$$

Find the missing number.
Start at 6. Count on to 9.

$$6 + \boxed{?} = 9$$

$$\boxed{?} = \underline{}$$

____ are in the .

Try It!

Choose the correct answer.

1. There are 6 in one can.
    There are 5 in another can.

    How many are there in all?

    10   11   12

    ⬭   ⬭   ⬭

2. There are 12 and 8 players.

    How many more are there than players?

    4   6   20

    ⬭   ⬭   ⬭

3. Ms. Wood buys 15 .

    She gives 8 to the children.

    How many does she have left?

    7   8   13

    ⬭   ⬭   ⬭

4. 13 players are on the field. 9 players are boys.

   How many players are girls?

   3   4   6

5. Ann buys 5 red balls, 3 green balls,
   and 2 purple balls.

   How many balls did Ann buy in all?

   10   13   15

6. The Tigers won 4 games in June.
   They won 3 games in July.
   They won 4 games in August.

   How many games did the Tigers win in all?

   8   11   12

7. Some 👓 are in the bag.
    Nate puts 6 more 👓 in the bag.
    Now there are 10 👓.

    How many 👓 were in the bag
    before Nate put 6 more in?

    _____

    Tell how you solved the problem.

    _____

    _____

    _____

    _____

    _____

Lesson 6 | Two-Digit Addition

Get the Idea

You can use models to help you add.

Example 1

$18 + 5 = $ ☐

Strategy Use models.

Step 1 Show 18. Add the 5 ones.

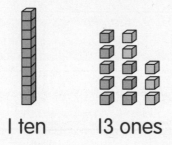

1 ten   13 ones

Step 2 Make 10 ones as 1 ten.

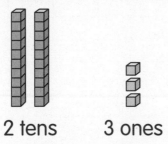

2 tens   3 ones

Step 3 Find the sum.
2 tens 3 ones = ☐23☐

Answer $18 + 5 = 23$

Example 2

$15 + 30 = \boxed{}$

Strategy Use a hundreds chart.

Step 1 Start at 15.

1	2	3	4	5	6	7	8	9	10
11	12	13	14	**15**	16	17	18	19	20
21	22	23	24	**25**	26	27	28	29	30
31	32	33	34	**35**	36	37	38	39	40
41	42	43	44	45	46	47	48	49	50
51	52	53	54	55	56	57	58	59	60
61	62	63	64	65	66	67	68	69	70
71	72	73	74	75	76	77	78	79	80
81	82	83	84	85	86	87	88	89	90
91	92	93	94	95	96	97	98	99	100

Step 2 Count 3 tens more.

15 ➜ 25, 35, 45

Answer $15 + 30 = 45$

You can use a place-value chart to help you add.

Example 3

$$27 + 40 = \boxed{}$$

Strategy Use a place-value chart.

Step 1 Add the ones.

7 ones + 0 ones = 7 ones

Tens	Ones
2	7
+ 4	0
	7

Step 2 Add the tens.

2 tens + 4 tens = 6 tens

Tens	Ones
2	7
+ 4	0
6	7

Step 3 Write the sum.

6 tens 7 ones = 67

Answer 27 + 40 = 67

Guided Practice

$29 + 5 = \boxed{}$

Use a hundreds chart.

1	2	3	4	5	6	7	8	9	10
11	12	13	14	15	16	17	18	19	20
21	22	23	24	25	26	27	28	29	30
31	32	33	34	35	36	37	38	39	40
41	42	43	44	45	46	47	48	49	50
51	52	53	54	55	56	57	58	59	60
61	62	63	64	65	66	67	68	69	70
71	72	73	74	75	76	77	78	79	80
81	82	83	84	85	86	87	88	89	90
91	92	93	94	95	96	97	98	99	100

Start at 29. Count on 5 ones.

29 ➜ 30 , _____ , _____ , _____ , _____

$29 + 5 =$ _____

Try It!

Add.

1. 34 + 30 = _____

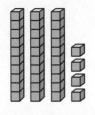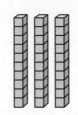

2. 16 + 6 = _____

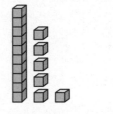

3. 36
   + 20
   _____

4. 25
   + 4
   _____

5. 36 + 3 = _____

6. 44 + 20 = _____

7.
Tens	Ones
2	3
+ 2	0

8.
Tens	Ones
5	5
+	3

9. Kim adds these numbers in different ways.

42 + 30 = ☐

Kim counts by tens. Show how she adds.

_____

Kim uses the chart. Show how she adds.

Tens	Ones
+	

Two-Digit Subtraction

Get the Idea

You can use models to help you subtract.

Example 1

$$\begin{array}{r} 70 \\ -\ 30 \\ \hline \end{array}$$

Strategy Use models.

Step 1 Show 70. Take away 30.

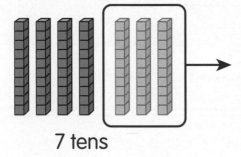

7 tens

Step 2 Count how many are left.

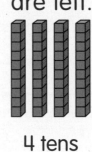

4 tens

Step 3 Write how many are left.

4 tens = 40

Answer

$$\begin{array}{r} 70 \\ -\ 30 \\ \hline 40 \end{array}$$

Addition and subtraction are opposites. Add the numbers to check your answer.

$$\begin{array}{r} 70 \\ -\ 30 \\ \hline 40 \end{array} \qquad \begin{array}{r} 40 \\ +\ 30 \\ \hline 70 \end{array}$$

Example 2

$$60 - 40 = \boxed{}$$

Strategy   Use a hundreds chart.

Step 1   Start at 60.

1	2	3	4	5	6	7	8	9	10
11	12	13	14	15	16	17	18	19	20
21	22	23	24	25	26	27	28	29	**30**
31	32	33	34	35	36	37	38	39	**40**
41	42	43	44	45	46	47	48	49	**50**
51	52	53	54	55	56	57	58	59	**60**
61	62	63	64	65	66	67	68	69	70
71	72	73	74	75	76	77	78	79	80
81	82	83	84	85	86	87	88	89	90
91	92	93	94	95	96	97	98	99	100

Step 2   Count back 4 tens.

60 → 50, 40, 30, 20

Answer   60 − 40 = 20

Guided Practice

50 − 30 = ☐

Use a place-value chart.
Subtract.

Tens	Ones
5	0
− 3	0

Check your answer.
Add.

Tens	Ones
2	0
+ 3	0

50 − 30 = _____

Operations and Algebraic Thinking

Try It!

Subtract.

1. 60 − 20 = _____

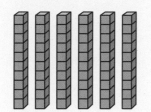

2.

Tens	Ones
7	0
− 5	0

3.

Tens	Ones
8	0
− 3	0

4.

Tens	Ones
9	0
− 3	0

5.   30
    − 20
    ‾‾‾‾

6.   40
    − 20
    ‾‾‾‾

7. Lee subtracts the numbers.

$$80 - 40 = \boxed{}$$

Lee uses the chart. Show how he subtracts.

Tens	Ones
−	

Lee checks his answer. Show how he adds.

Tens	Ones
+	

Operations sund Algebraic Thinking

Glossary

E

equal sign (=) (Page 16)

$$7 + 4 = 11$$

equal sign

equation a number sentence with an equal sign (=) (Page 16)

$$4 + 6 = 10$$
$$9 - 2 = 7$$
$$5 + 3 = 3 + 5$$

F

fact family a group of related facts (Page 4)

$$4 + 5 = 9$$
$$5 + 4 = 9$$
$$9 - 5 = 4$$
$$9 - 4 = 5$$

Math Tools

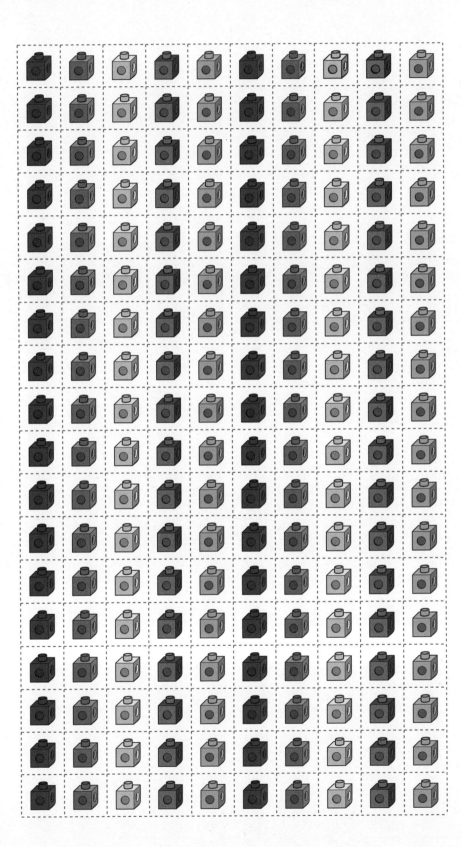

Math Tools

Math Tools

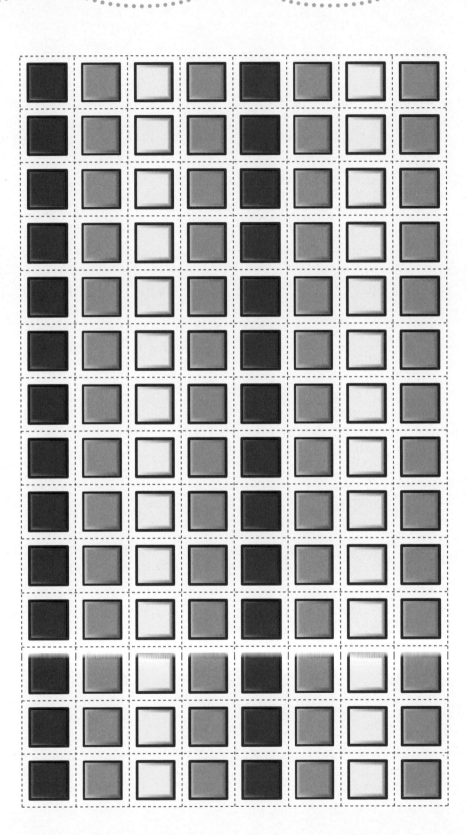

Math Tools

Math Tools

1	2	3	4	5	6	7	8	9	10
11	12	13	14	15	16	17	18	19	20
21	22	23	24	25	26	27	28	29	30
31	32	33	34	35	36	37	38	39	40
41	42	43	44	45	46	47	48	49	50
51	52	53	54	55	56	57	58	59	60
61	62	63	64	65	66	67	68	69	70
71	72	73	74	75	76	77	78	79	80
81	82	83	84	85	86	87	88	89	90
91	92	93	94	95	96	97	98	99	100

Math Tools

Tens	Ones

+

Tens	Ones

−

Tens	Ones

+

Tens	Ones

−

Tens	Ones

+

Tens	Ones

−